Gravitational Interactions

Alyxx Meléndez

Consultants

Samantha M. Thompson
Curator of Science and Technology
National Air and Space Museum

Cheryl Lane, M.Ed.
Seventh Grade Science Teacher
Chino Valley Unified School District

Michelle Wertman, M.S.Ed.
Literacy Specialist
New York City Public Schools

Publishing Credits

Rachelle Cracchiolo, M.S.Ed., *Publisher*
Emily R. Smith, M.A.Ed., *SVP of Content Development*
Véronique Bos, *VP of Creative*
Dani Neiley, *Editor*
Robin Erickson, *Senior Art Director*

Smithsonian Enterprises

Avery Naughton, *Licensing Coordinator*
Paige Towler, *Editorial Lead*
Jill Corcoran, *Senior Director, Licensed Publishing*
Brigid Ferraro, *Vice President of New Business and Licensing*
Carol LeBlanc, *President*

Image Credits: p. 9 Smithsonian Magazine; p. 11 NASA; p. 15 Smithsonian
National Air and Space Museum; pp. 15–18 NASA; p. 19 courtesy Author Woods;
p. 19 NASA; p. 20 Smithsonian National Air and Space Museum; pp. 21, 23, 32 NASA;
all other images iStock and/or Shutterstock, or in the public domain

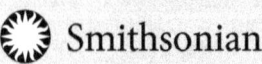
Smithsonian

5482 Argosy Avenue
Huntington Beach, CA 92649
www.tcmpub.com
ISBN 979-8-7659-6876-5
© 2024 Teacher Created Materials, Inc.

Table of Contents

Planetary Pull

Gravity is a force that pulls all objects together. Objects with greater mass have stronger gravitational pulls than objects with less mass. Mass is the amount of matter, or "stuff," in an object. Objects that are very **dense**, such as neutron stars and black holes, also have strong gravitational forces.

So why don't large objects, such as blue whales and skyscrapers, go flying toward one another all the time? The answer lies in their size. Whales and skyscrapers certainly seem massive to humans. But in the grand scheme of things, they are really quite small. From a planet's point of view, even the tallest building on Earth is still miniscule. Only objects as big as stars, moons, and planets are sizeable enough to have gravity that a person can feel.

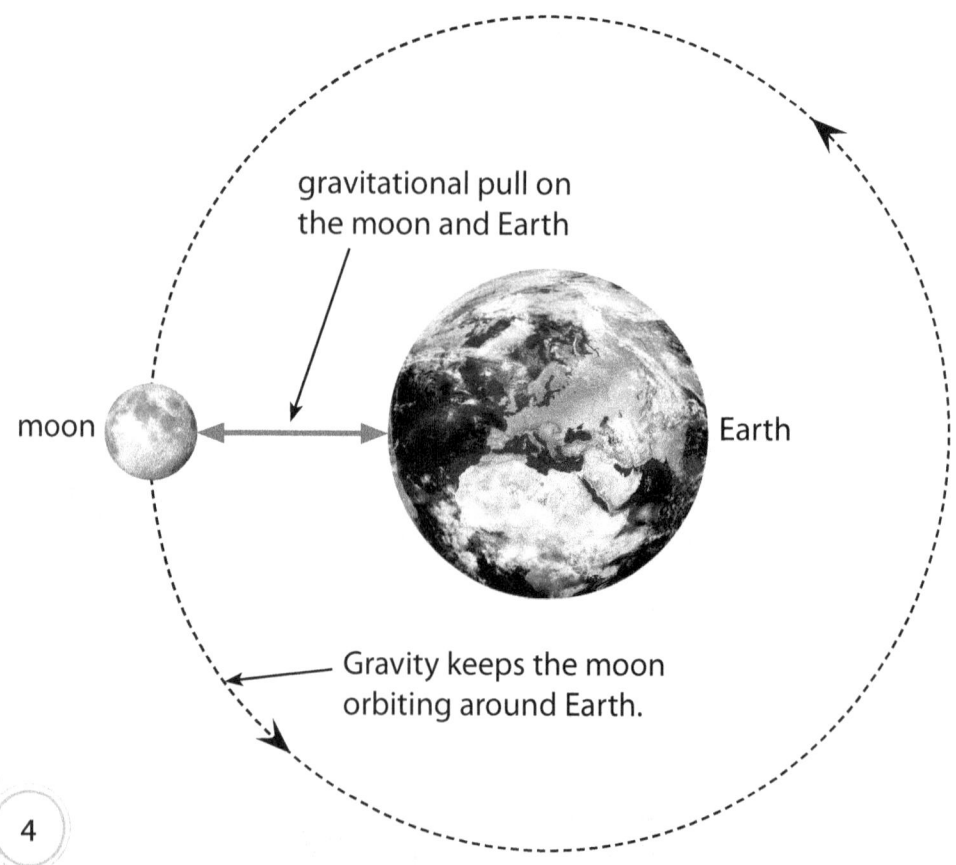

gravitational pull on the moon and Earth

moon

Earth

Gravity keeps the moon orbiting around Earth.

Gravity is what keeps planets in orbit. Planets are pulled toward one another's centers of mass. The objects with the most mass are able to affect the orbits of smaller objects. In our solar system, that large object is the sun. The sun is so massive, it could fit more than a million Earths inside it! That kind of mass allows the sun to send eight planets plus Pluto (a dwarf planet) spinning around it.

Gravity has a profound effect on everything touching Earth's surface. In fact, gravity exists everywhere—and it has different effects depending on where you are.

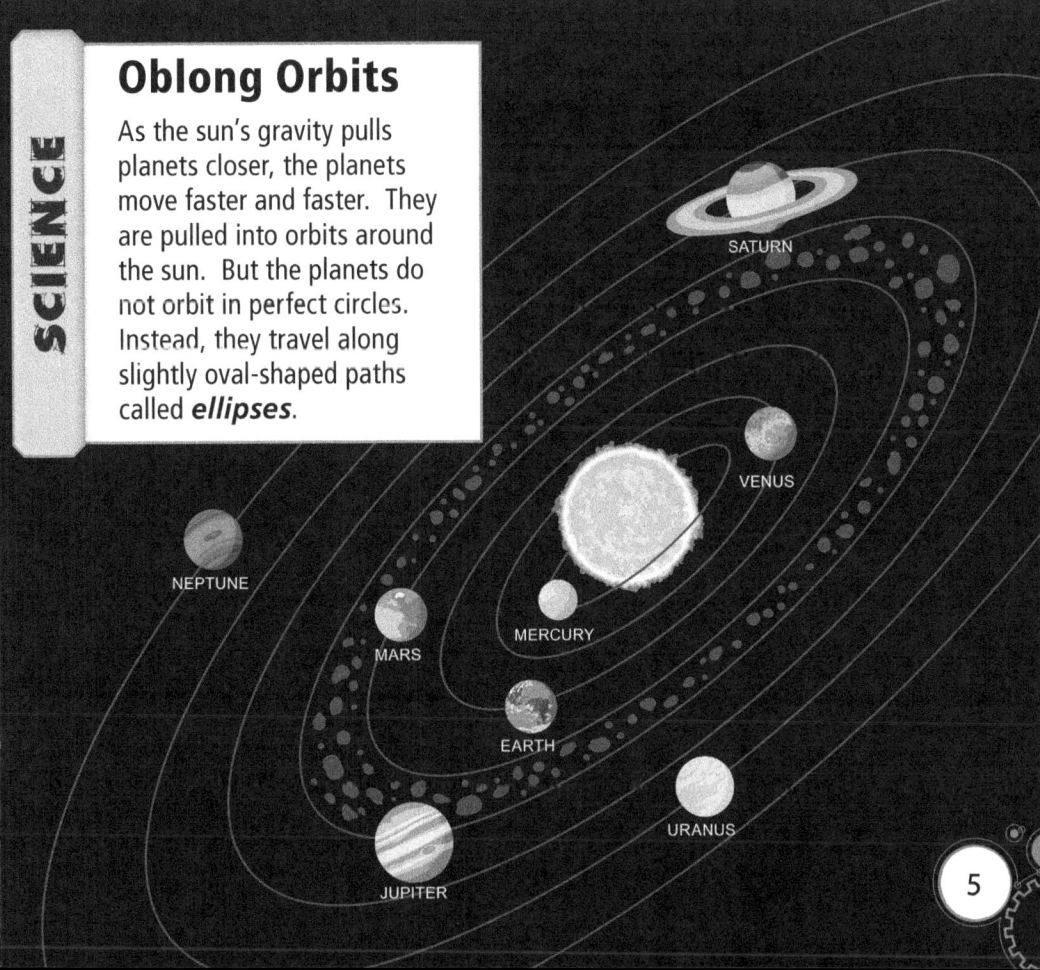

SCIENCE

Oblong Orbits

As the sun's gravity pulls planets closer, the planets move faster and faster. They are pulled into orbits around the sun. But the planets do not orbit in perfect circles. Instead, they travel along slightly oval-shaped paths called **ellipses**.

SATURN

VENUS

NEPTUNE

MERCURY

MARS

EARTH

URANUS

JUPITER

Gravity on Earth

Gravity is measured by how fast an object **accelerates** when it falls through the sky. The acceleration of gravity on Earth is 32 feet (9.8 meters) per second squared. This means that for every second an object spends falling, its speed increases by that amount every second. To put this into perspective, imagine dropping a watermelon from a second-story balcony. As it hurtles toward the ground, it gets faster and faster. Then, it hits with a SPLAT! Dropping it from a lower height would have a much lower gravitational effect. The watermelon would end up bruised instead of split open.

While gravity pulls down toward Earth's surface, air molecules push up against falling objects. Airborne objects, such as parachutes, use air to resist Earth's gravitational pull. Parachutes slow the force of gravity by taking advantage of air resistance. Parachutes give skydivers a larger surface area for air molecules to press against. This helps them slow down as they fall.

In comparison, birds have an even greater command over air resistance. Their wings move air the same way a swimmer's hands move water. Birds flap their wings to push down air molecules faster than gravity can pull their bodies down. To prepare for the next wing flap, they slice their wings forward through the air. This motion lets birds cut through air resistance.

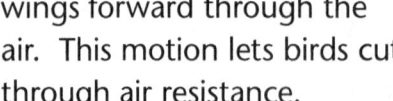

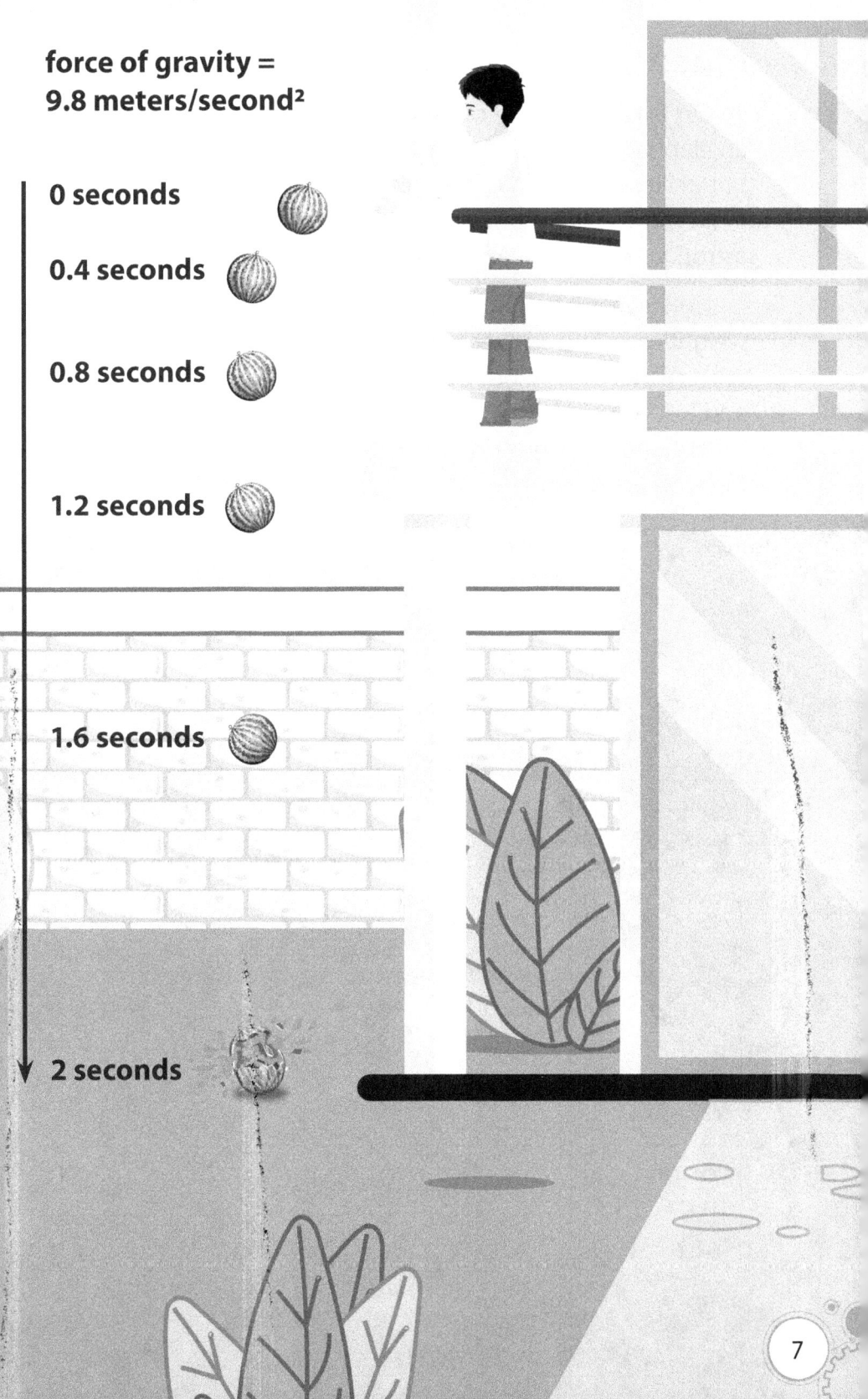

**force of gravity =
9.8 meters/second²**

0 seconds

0.4 seconds

0.8 seconds

1.2 seconds

1.6 seconds

2 seconds

Mass versus Weight

On Earth, anybody can lift a butterfly. But nobody can lift an elephant—not without machinery, at least. But in space, butterflies and elephants would appear to weigh the same. In fact, they would both float as if they were weightless. An astronaut could move both animals with a single tap.

How can something as large as an elephant seem weightless? The answer has to do with gravity. No matter what planet they're on, elephants always have lots of mass, and butterflies always have much less mass. Weight, on the other hand, depends on gravity. It is the measure of gravity's force on an object.

Every **celestial body** has its own gravitational pull. Celestial bodies with lots of mass, such as Jupiter, have stronger gravity. Celestial bodies with less mass, such as the moon, have weak gravity. And in outer space, far away from any celestial body, there's hardly any gravity at all. If an astronaut stepped on a scale in space, their weight would read "zero." So, to measure mass in space, a special scale is needed. This scale measures objects by moving them with a certain amount of applied force. Then, the scale shows how much the object accelerated. These special scales can use mathematical properties to calculate an object's mass!

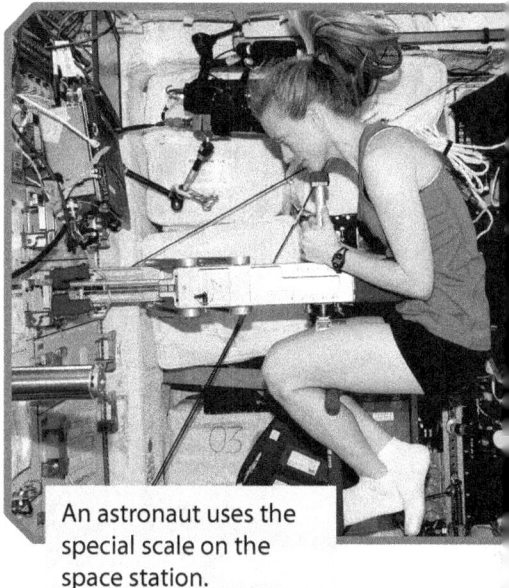

An astronaut uses the special scale on the space station.

MATHEMATICS

Newton's Laws

Isaac Newton was an English mathematician and physicist. He had a lot of important ideas about gravity. In 1687, he published an idea to explain the motions of the planets and their moons. Today, that idea is known as Newton's law of gravitation.

Gravity on the Moon

When two celestial bodies pass, the one with more gravity can pull the other into its orbit. Since gravity depends on mass, the more massive body will pull in the smaller one. This means moons are always less massive than the planets they orbit. Earth's single moon is no exception. It has just over one percent of Earth's mass. Therefore, its gravity is far weaker than Earth's. The moon's gravity is about 17 percent of Earth's gravity.

To understand how gravity affects objects on the moon, think about the watermelon and the balcony again. If an astronaut on the moon dropped the same watermelon off the same balcony, it would not fall the way it did on Earth. This is because the acceleration of gravity on the moon is much lower compared to Earth. The acceleration of gravity on the moon is about 5 feet (1.6 meters) per second squared.

When dropped, the watermelon would gently drift toward the moon's surface. Its fall would be slower than that of an Earth-bound watermelon. The watermelon would come to rest on the lunar dust with a soft *thump*. The rind might bruise, but it would not split open completely.

When astronauts visit the moon, the **reduction** in gravity changes the way they move. Every footstep takes a longer time to reach the ground than it does on Earth. If astronauts try to jump, they appear almost as if they were filmed in slow motion. They move very differently from how they move on Earth!

FUN FACT

The first successful lunar landing mission was Apollo 11. Neil Armstrong and Buzz Aldrin were the first humans to set foot on the moon. So far, only 12 people have walked on the moon!

astronaut John W. Young during the Apollo 16 mission

Earth and the Moon

The moon's gravity has an effect on Earth's oceans. When high tide approaches, beachgoers need to be careful of the incoming water. The thing responsible for those huge tidal forces is the moon. At the same time as Earth's gravity pulls on the moon, the moon's gravity pulls back against Earth. So, when the moon faces one side of Earth, the moon pulls all the water on that side of Earth toward it. This is where the tide is highest. There will also be a lower high tide on the opposite side of Earth—the side farthest from the moon.

At high tide, water typically covers most of a shore.

Low tide often gives beachgoers more room to walk and play in the sand.

On the two sides **perpendicular** to the moon, the tide will be low. It takes about 24 hours and 50 minutes for the moon to move between tide cycles. Beaches normally experience two high tides and two low tides each day.

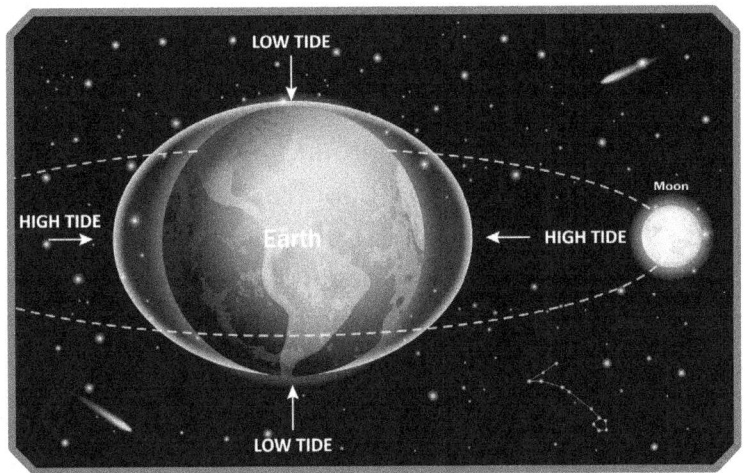

If the moon suddenly disappeared one day, the tides would be only about a third as big as they are now. Tidal forces would still exist, but they'd only be controlled by the gravitational pull of the sun. Since the sun is more than 94 million miles (150 million kilometers) from Earth, its gravity has a weaker effect on Earth's oceans. Thrill-seeking surfers have the moon (as well as wind) to thank for the giant waves they ride!

Gravity in Space

Astronauts on the International Space Station, or ISS, do research in outer space. They share videos from inside the station. In these videos, they float across the screen, and their hair hovers around their heads. It may seem as though gravity isn't acting on them whatsoever. But this isn't exactly true. Gravity causes every object in space to pull every other object toward it. That includes the ISS. This space station orbits Earth, whizzing around the planet 248 miles (400 kilometers) away. So, although astronauts are farther than usual from Earth, a tiny bit of gravity still pulls on them. Astronauts experience **microgravity** in space, which makes people or objects appear to be weightless.

International Space Station

Sometimes, astronauts leave the ISS to do work outside the station. They might do experiments or perform maintenance on the ISS. But before they can go on **spacewalks**, astronauts need to put on their pressurized spacesuits. They spend a few hours breathing pure oxygen. This removes all the nitrogen from their blood.

Astronauts work together on a spacewalk.

Without this preparation, astronauts could get gas bubbles in their bodies and experience health problems.

Normally, spacewalking astronauts stay **tethered** to their spacecraft. This keeps them from floating away. On some space flights, astronauts did not use a tether. Instead, they used the Manned Maneuvering Unit, or MMU. This device is like a jet-powered backpack. It gives astronauts control over their movement in a low-gravity environment.

Real-Life Jetpack

Gyroscopes help the MMU point in the right direction. Then, thrusters push out compressed nitrogen. This is similar to the compressed gas that helps push whipped cream out of a can. The MMU **propels** astronauts in any direction they choose.

gyroscope

Super-Strong Gravity

Black holes are an example of extreme gravity. They can be created from a supernova, which is a dying star. As the dying star collapses, it creates a teeny-tiny space known as a black hole. There, gravity is so strong that not even light can get out. Black holes also pull light into themselves, so they are invisible to telescopes. But their gravity can be strong enough to affect the stars around them. Scientists study those stars to find out where black holes are.

In our solar system, the only star is the sun, and it is too small to become a black hole. In fact, all black holes that scientists know of are **light-years** away from Earth. Using special tools, scientists have been able to study what happens when a black hole's super-strong gravity pulls in stars. The stars become stretched so much that they are torn apart. Some scientists have wondered if the same thing would happen if a human got near a black hole.

Here's what scientists think might happen: As a person approaches a black hole, their body would become stretched out vertically and pushed in horizontally. Gravity would turn them into a long, thin noodle of the person they once were. This is called **spaghettification**. Once someone has been spaghettified and sucked into a black hole, no one knows what happens next.

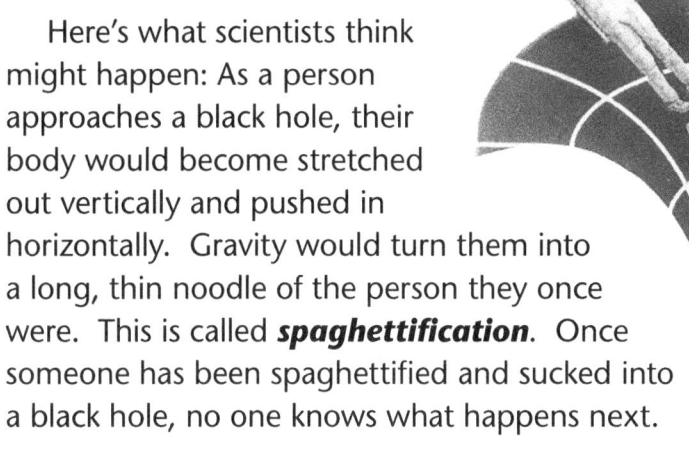

FUN FACT

Astronomers have to use special technology to find black holes. NASA's Chandra X-ray Observatory is one of the tools they use. As matter is pulled toward black holes, it becomes super hot. It is heated to millions of degrees! These hot spots glow in X-rays.

Chandra (below) took this image of two black holes.

Astronaut Life Hacks

Microgravity affects all parts of an astronaut's day-to-day life in space. So, astronauts need special tools to help them do everyday activities. For example, in space, water doesn't stay in a bathtub or fall toward a drain. Instead, it floats around and is hard to control. So, how do astronauts keep clean? They fill up and use their reusable plastic water bags. Bubbles of water float out of the water bags whenever they are squeezed. Astronauts grab these bubbles out of the air and use them to soak toothbrushes, washcloths, and hair.

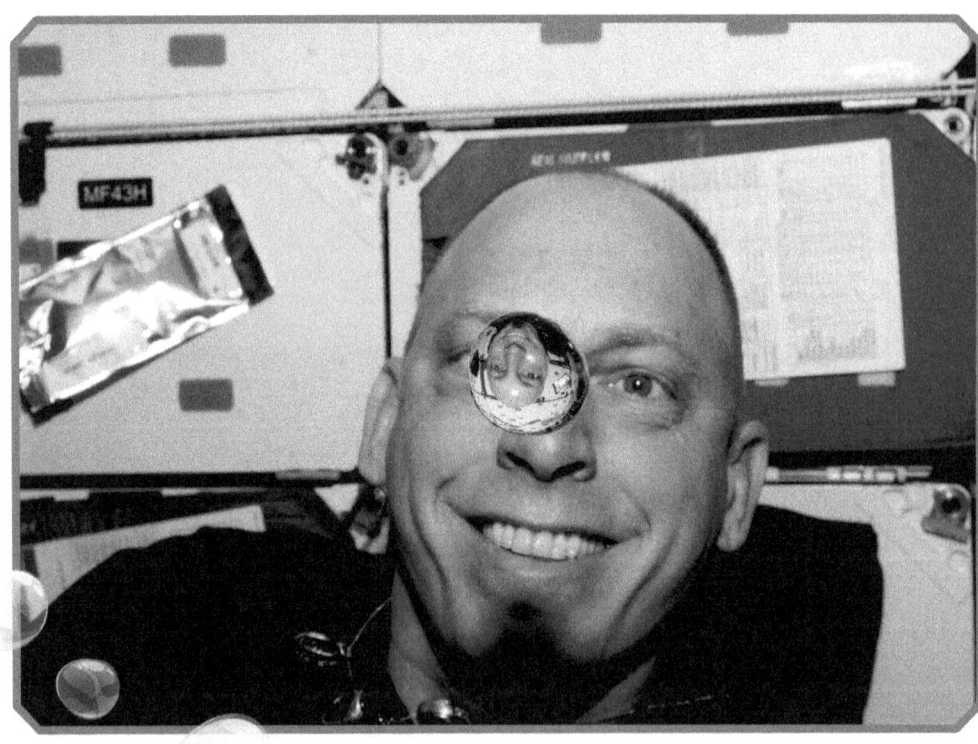

Each spacecraft has special bathrooms for astronauts. Two toilets can be found on board. One is for liquid and another is for solid waste. Both toilets are equipped with suction to prevent the floating waste from causing any messy mishaps. Liquid waste is recycled back into drinking water, and solid waste is packaged and burned.

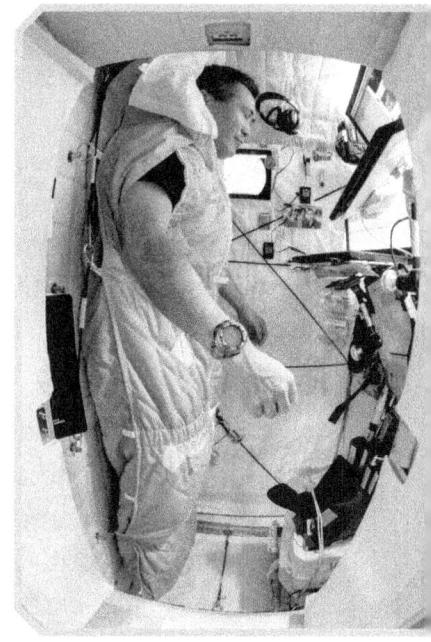

Sleeping astronauts may look a bit uncomfortable strapped into their beds. However, it would be much harder to catch some shut-eye if they were floating around and bouncing off the walls! When it's bedtime, astronauts zip themselves into anchored sleeping bags. They can sleep in any direction thanks to microgravity. When asleep, they look like they're playing invisible pianos. Microgravity isn't strong enough to pull their arms down to their sides!

ARTS

Space Sculpture

Astronauts wanted to study how microgravity would affect a sculpture. So, they sent artist Arthur Woods's sculpture *Cosmic Dancer* to the Mir Space Station. In space, the sculpture spun so it could be viewed from all angles. Astronauts felt inspired to spin and dance along!

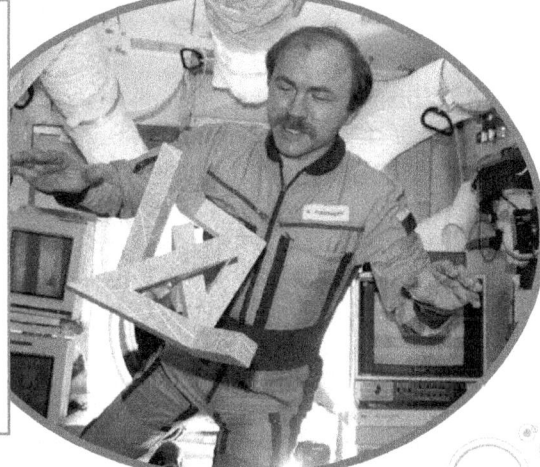

19

Changes in Space

In space, the human body faces some uncomfortable changes without the gravity it is used to on Earth. When astronauts go to space, they first notice a shift in their blood circulation. Since there is hardly any gravity to push blood down to their feet, blood rises to their heads instead. Their faces can swell up, and their legs start to look skinny. Astronauts sometimes call this the "puffy head bird legs" **phenomenon**. All that extra fluid in astronauts' heads can put pressure on their eyes. Their eyeballs begin to flatten, which can severely damage their vision.

In space, astronauts hardly need to move a muscle. They can sit, stand, and somersault in midair with minimal effort. This is quite different from Earth, where just staying awake can be a workout! For example, eyelid muscles have to work harder against Earth's gravity to stay open. But in space, astronauts' muscles have to work way less. Plus, their bones don't need to make any extra bone cells to help hold up their bodies. For this reason, it is important for astronauts to get at least an hour of daily exercise. If they don't, their bones and muscles will **deteriorate**. Even their heart muscles will weaken, since it takes less effort to pump blood.

FUN FACT

The Apollo 7 mission was the first time astronauts had enough room on their spacecraft to exercise. During the mission, astronauts used the Exer-Genie Exerciser. It was a small rope that went around a metal shaft. Astronauts would hook it to the wall of the spacecraft and pull on the rope to cause friction. The tool helped them stretch and exercise their muscles.

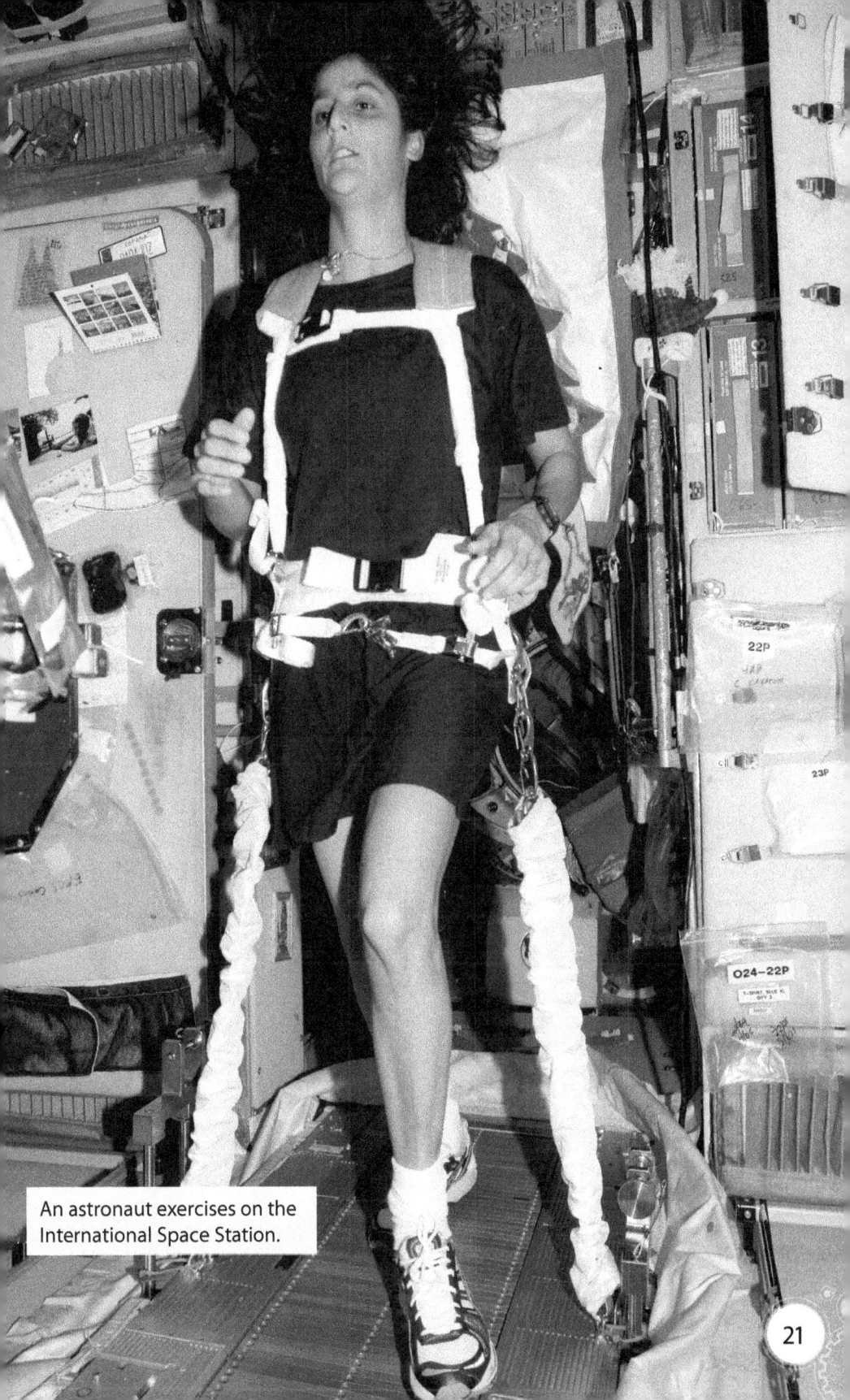

An astronaut exercises on the International Space Station.

Zero-G on Earth?

Although there is no real way to escape gravity on Earth, there is a way to fake microgravity. Let's return to the watermelon and the balcony one last time. Imagine that this time, a caterpillar has crawled inside the melon. The caterpillar ate the whole watermelon, and it's inside the empty rind. What happens to the caterpillar when someone drops the watermelon off a second-story balcony? For a moment, the caterpillar would feel weightless and float around inside the rind. What this bug would experience is **artificial** microgravity.

Airplanes can also simulate microgravity. They do this by flying up and down in a **parabolic** path. A parabola is a U-shaped curve. On the way up, people in the plane feel about twice as heavy. As the plane reaches the top of the arc, people begin to experience weightlessness as they enter free fall. They experience microgravity for about 20–30 seconds. These types of flights have been used for research purposes to study the effects of microgravity.

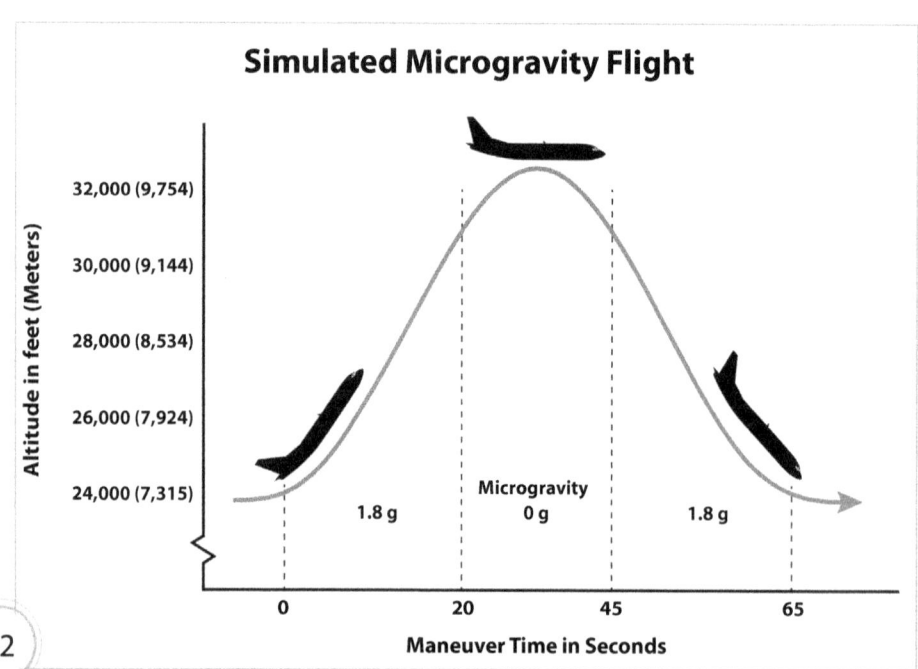

Simulated Microgravity Flight

Three pilots are necessary to control parabolic motion. The first pilot points the airplane's nose up or down. This is the hardest job because the airplane must point at specific angles for specific amounts of time. The second pilot keeps the wings horizontal and stable, no matter how the rest of the plane moves. The third pilot controls the plane's speed.

This flight simulates the effects of microgravity.

G-Force

Daredevils don't need to put on spacesuits to test the force of gravity. They can just visit their local amusement park and hop on a roller coaster! Gravity is what makes these clickety-clacking carts so exciting to ride.

As a cart goes up the first hill, it accelerates upward against the force of gravity. Acceleration is measured in **g-force**, or g's. At the lowest level, 1 g is equal to the normal push of gravity on Earth. Any g-force greater than 1 g feels heavier than Earth's normal gravity. When riders accelerate in the opposite direction from gravity, they feel heavy, as though the air itself is pushing them down. This is known as positive g-force. Coaster riders can also feel positive g-force during sharp turns. This makes their blood rush into their feet. If the force is too strong, it can make people feel dizzy or faint.

At the hill's peak, the cart starts hurtling downward. It accelerates in the same direction as gravity. Riders will feel weightless for a moment, just like passengers on a microgravity airplane. This is known as negative g-force. Negative g's can pull riders out of their seats. This is why modern coasters strap their riders down with seat belts and harnesses.

Coaster Chaos

One of the first looped roller coasters, the Flip Flap Railway, had a circular loop. This shape put a lot of g-force on its passengers, which caused whiplash and injury. Some people even fell out! Today, roller coaster loops are not perfect circles. They are shaped like ellipses, which makes for much safer riding.

Gravity Is Everywhere

No matter where an object is on Earth or in space, gravity acts upon it. Gravity can be affected by the distance between objects and the size of objects. This force causes all objects to move faster and faster toward one another. But of course, only the biggest or densest objects have noticeable gravitational pulls.

Gravity looks different in different environments. On Earth, humans use gravity to their advantage. Devices such as parachutes and airplanes allow humans to soar through the air. Winged animals, such as birds and bats, conquer gravity all on their own. Gravity between Earth, the sun, and the moon controls tidal forces. On the moon, astronauts find themselves stepping and hopping much slower. And in outer space, astronauts use swimming motions to get around. They need special tools to help them do everyday activities.

Gravity can be felt anywhere in the universe. Each planet and moon has its own special gravitational force. Even astronauts far out in outer space feel some kind of gravity. This universal force truly does bring all people and things together!

STEAM CHALLENGE

Define the Problem

People have long been trying to defy gravity and use it for thrills. You can see this clearly at amusement parks. An amusement park near you is planning to put in a new roller coaster. To engage young engineers, they are having a contest. Your task is to engineer a model section of a roller coaster that a marble can safely "ride." If it works, your design could be chosen to be included in a part of the new roller coaster.

 **Constraints:** You may only use the materials provided to you.

 **Criteria:** Your model roller coaster section must stand up straight. It must have one upside-down loop in it, and the marble must stay on the track from beginning to end.

Research and Brainstorm

How will gravity help your marble make its way along your roller coaster section? What other forces will be at work? How will you keep your roller coaster section upright? How high will the start of your roller coaster section be? What shape and size will work best for the upside-down loop?

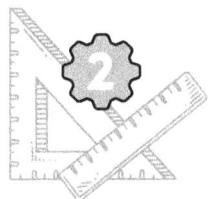

Design and Build

Sketch two or more designs for your marble roller coaster loop. Label the parts and the materials. Choose the design you think will work best. Then, build your roller coaster section.

Test and Improve

Place your marble at the beginning of your roller coaster section, and let it go. Does it stay on the track? What about your design is working well? What new goals can you set for your design? Can you add a second loop or a new section? How can you improve the look of your design? Modify your design, and rebuild it as needed. Reassess how well it meets the criteria.

Reflect and Share

What part of this challenge are you most proud of? What problems did you encounter and how did you resolve them? What challenges would you face if you were engineering a roller coaster for the surface of the moon?

Glossary

accelerate—increase in speed; move faster and faster

artificial—humanmade

celestial body—a singular object in space, such as a star, planet, moon, or asteroid

dense—having a high mass per unit volume

deteriorate—to become worse over time

ellipses—ovals

g-force—the force of gravity or acceleration on a body

gyroscopes—wheels or discs that spin rapidly and can turn freely in any direction

light-years—units of length equal to how far light can travel in a year; about 5.88 trillion miles (9.46 trillion kilometers)

microgravity—the condition of appearing to be weightless in the near absence of gravity

parabolic—having a form like a symmetrical, bowl-shaped curve or wave

perpendicular—forming a 90-degree angle with another line or object

phenomenon—an interesting or unusual fact or event that people can study

propels—pushes forward

reduction—the act or process of making something smaller or weaker

spacewalks—periods of movement in space outside a spacecraft by astronauts

spaghettification—vertical stretching and horizontal compression caused by strong gravity

tethered—tied to a stable object to stop the other object from moving too far

Index

CAREER ADVICE

from Smithsonian

Do you dream of becoming an astronaut?

Here are some tips to keep in mind for the future.

"Astronauts usually have other jobs first. They may be teachers, pilots, engineers, scientists, or doctors. Focus on what you enjoy, and try to learn as much math and science as you can, too."

– *Samantha M. Thompson, Curator of Science and Technology, National Air and Space Museum*

"Be curious and learn as much as you can. I'm a firm believer that all kinds of knowledge can turn out to be useful in any kind of career."

– *Margaret Weitekamp, Chair of Space History and Curator, National Air and Space Museum*